S0-ACS-542

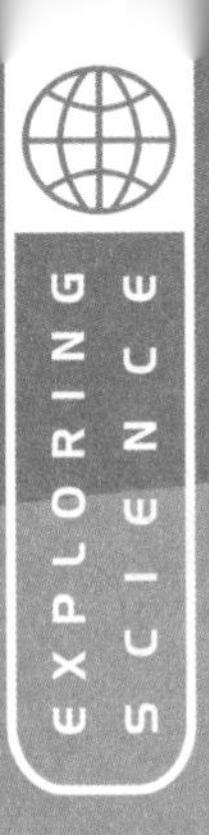

MAJOR ORGANS

SUSTAINING LIFE

BY SHARON KATZ COOPER

Content Adviser: Debra Carlson, Ph.D., Associate Professor, Normandale Community College, Bloomington, Minnesota

Science Adviser: Terrence E. Young Jr., M.Ed., M.L.S., Jefferson Parish (Louisiana) Public School System

Reading Adviser: Rosemary G. Palmer, Ph.D., Department of Literacy, College of Education, Boise State University

Compass Point Books • Minneapolis, Minnesota

Compass Point Books • 3109 West 50th Street, #115 • Minneapolis, MN 55410

Visit Compass Point Books on the Internet at *www.compasspointbooks.com*
or e-mail your request to *custserv@compasspointbooks.com*

Photographs ©: Alfred Pasieka/Peter Arnold, cover, 5; Science Photo Library/Photo Researchers, 4, 9, 17, 41; Alexander Tsiaras/Photo Researchers, 7; Anatomical Travelogue/Photo Researchers, 8, 18, 21, 23, 30, 31, 32, 39; BSIP/Photo Researchers, 10; Sophie Jacopin/Photo Researchers, 11; Perennon Nuridsany/Photo Researchers, 12; AJ Photo/Photo Researchers Inc., 13; Susumu Nishinaga/Photo Researchers, 14, 40; Asa Thoresen/Photo Researchers Inc., 16; Coneyl Jay/Photo Researchers, 24; Morris Huberland/Photo Researchers Inc., 25; CNRI/Photo Researchers, 29; David Musher/Photo Researchers, 33; Royality Free/Corbis, 34; Steve Gschmeissner/Photo Researchers, 35; Lester V. Bergman/Corbis, 36; Visuals Unlimited/Corbis, 37; Sheila Terry/Photo Researchers, 38; Dennis Wilson/Corbis, 42; Michael Hagedorn/Zefa/Corbis, 43; Mehau Kulyk/Photo Researchers, 44; Joubert/Phanie/Photo Researchers, 46.

Editor: Anthony Wacholtz
Designer/Page Production: Bobbie Nuytten
Photo Researcher: Lori Bye
Cartographer: XNR Productions, Inc.
Illustrators: Eric Hoffmann

Art Director: Jaime Martens
Creative Director: Keith Griffin
Editorial Director: Carol Jones
Managing Editor: Catherine Neitge

Library of Congress Cataloging-in-Publication Data
Cooper, Sharon Katz.
Major organs : sustaining life / by Sharon Katz Cooper.
p. cm. — (Exploring science)
Includes bibliographical references and index.
ISBN-13: 978-0-7565-1959-9 (library binding)
ISBN-10: 0-7565-1959-4 (library binding)
ISBN-13: 978-0-7565-1965-0 (paperback)
ISBN-10: 0-7565-1965-9 (paperback)
1. Body, Human—Juvenile literature. 2. Human anatomy—Juvenile literature. 3. Human physiology—Juvenile literature. I. Title. II. Series.
QP37.C7892 2007
612—dc22 2006027047

About the Author

Sharon Katz Cooper is a writer and science educator. She enjoys writing about science and social studies topics for children and young adults. She lives in Fairfax, Virginia, with her husband, Jason, and son, Reuven.

TABLE OF CONTENTS

CHAPTER ONE

What Is an Organ?

YOUR HEART BEATS as you run to catch a bus. Your muscles stretch as you prepare for a track meet. Your stomach grumbles as you wonder when dinner will be ready. These are all signs of your body's major organs at work.

The body is composed of various kinds of cells, the smallest units of life. Each cell has a particular function. When groups of similar cells work together to perform a certain function, such as protection or communication, they form a tissue. There are four major types of tissues—connective, epithelial, muscle, and nerve. Connective tissues link various parts of the body or hold parts of the body in place. Epithelial tissues are made of tightly packed cells that form protective barriers against outside threats or between different parts of the body. Muscle and nerve tissues, as their names suggest, make up muscles and nerves.

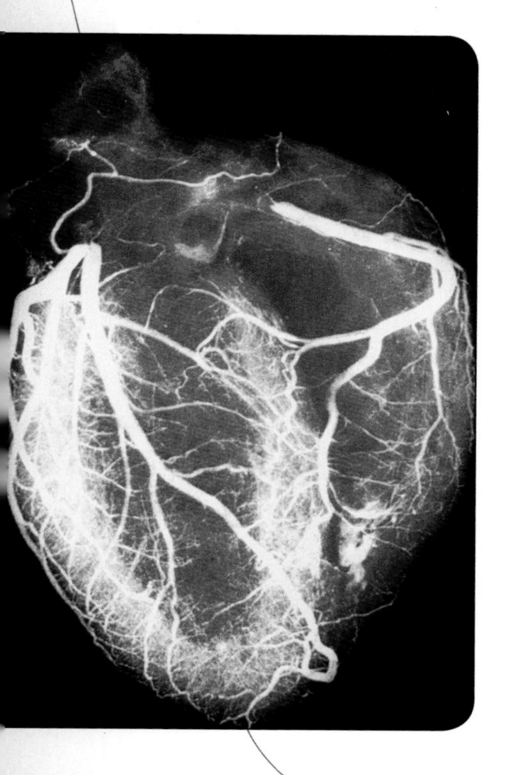

The arteries within a heart can be seen in an angiogram, which is an X-ray picture made after injecting the blood vessels with a dye to make them visible.

A group of two or more tissues that work together to perform a function is called an organ. For example, the stomach and heart are organs that are made of epithelial, connective, and muscle tissues. The stomach processes food for digestion, and the heart pumps blood throughout the body.

Groups of organs that work together make up an organ system. The nervous system is one example of a human body system. Its major organs are the brain, spinal cord, and nerves. The major organs of the digestive system include the esophagus, stomach, and intestines.

Each system has a particular function to carry out. However, each system also interacts with other systems. For example, when you eat, the digestive system works to break apart food into nutrients the body can use. At the same time, the circulatory system moves those nutrients through blood vessels to all parts of the body, the respiratory system provides oxygen to the stomach and intestine muscles, and the urinary system recycles fluids and disposes of waste.

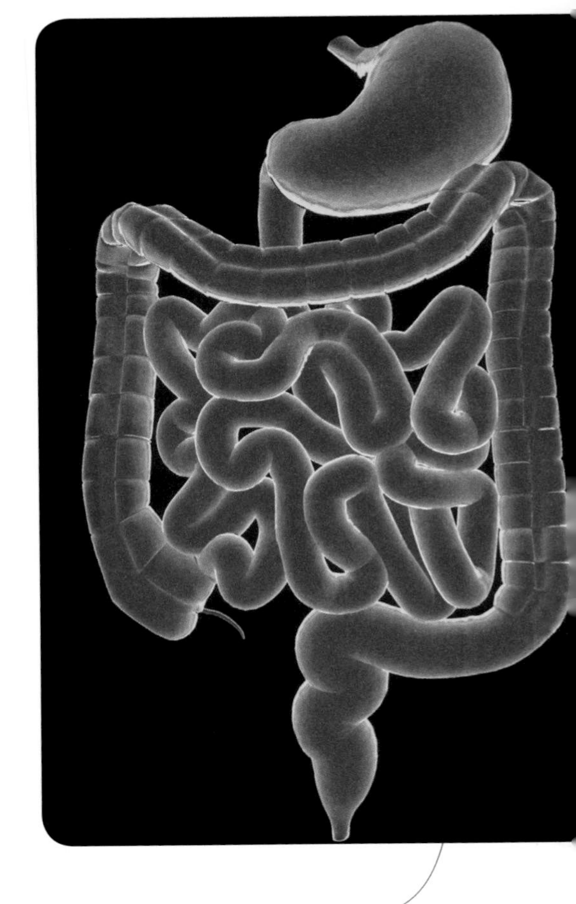

The organs of the digestive system are easily distinguished in a computer-generated graphic.

Organ Transplants

Organs usually function well unless a person is injured or has a serious disease in a particular organ. If the organ is not able to perform its job and it cannot be fixed, an organ transplant may be an option. An organ transplant is a major surgery in which doctors replace a damaged or diseased organ with a healthy organ from a donor.

Most donated organs—including hearts, livers, and intestines—come from people who have recently died. These donors are usually accident victims whose organs were otherwise healthy. Organs that can be donated by living people include kidneys and parts of the liver and pancreas. After a kidney donation, the remaining kidney is usually able to perform the work of two. People who donate parts of their healthy liver or pancreas usually have few problems afterward.

The United States manages organ transplants nationally through the Organ Procurement and Transplantation Network. This network keeps a computerized list of people who need an organ donation. When an organ becomes available, doctors enter information about the donor into a computer. A software program searches for potential matches, analyzing factors such as the blood type, tissue type, and organ size. It also looks at the amount of time each person has been waiting for an organ, the urgency of each situation, and the distance

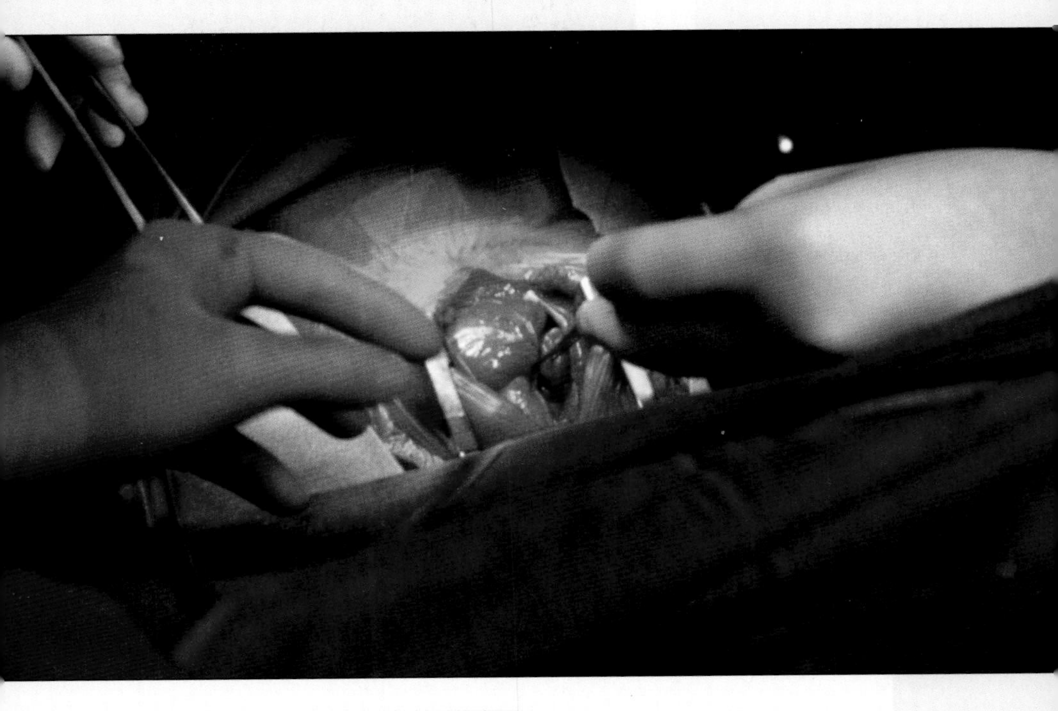

between the donor and each possible recipient. The computer then produces a ranked list of potential recipients. After reviewing the list, a doctor will call the highest-ranked patient, make arrangements to transport the organ, and schedule the surgery.

Each day, about 74 people in the United States receive organ transplants. At the same time, however, approximately 91,000 people are waiting for transplants.

A machine took over the function of the heart and lungs of a young boy while he received a heart transplant.

CHAPTER TWO

Supportive and Protective Organs

THE LARGEST ORGAN of your body is the skin. Skin is an elastic, flexible organ, and it is composed of all four kinds of tissues. The skin's most important function is protection. It responds to many kinds of stimuli, such as cold, heat, touch, pressure, and pain. It also helps the body to maintain a steady temperature through its sweat glands and hair.

The epidermis is the outer covering of the skin. The top level of the epidermis is composed of 25 to 30 individual layers of flattened, dead cells. The body constantly sheds these cells. But even dead cells can be important. They help protect the skin's inner layers. Directly beneath the dead layers are living cells that grow and replace the

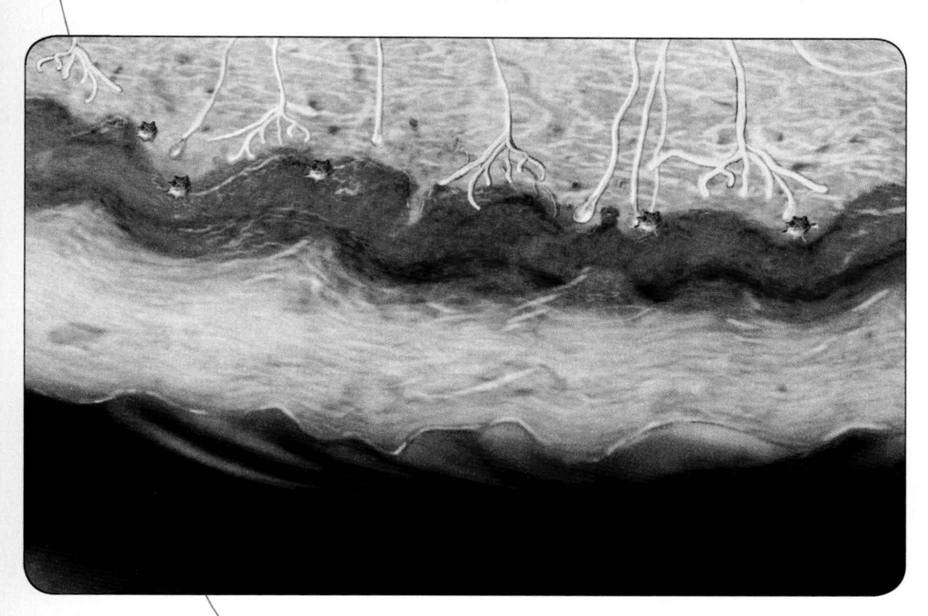

A cross-section of a human finger reveals the epidermis, dermis, blood vessels, and nerve receptors.

dead ones. These living cells contain a substance called melanin that provides color to the skin and protects it from the sun's radiation. These newly formed cells constantly push up toward the surface and eventually die as well.

Underneath the epidermis is a thicker layer of skin called the dermis. It contains blood vessels, nerves, and sweat and oil glands. It also contains fat deposits to cushion and insulate the body. Hair grows out of hair follicles, which are deep indentations in the dermis, and extends through the epidermis. These hair follicles have their own blood vessels and nerves, and they are attached to muscle tissue.

Eyelashes are anchored into hair follicles in the skin.

Acne

Acne is a skin problem that affects most people sometime in their lives. It is most common during puberty, when increased hormones make the body produce more oil.

Acne starts when the epidermis produces extra cells around a hair follicle. These cells stick to each other in a large clump, mix with oil, and form a blockage in the follicle. When the follicle becomes blocked, bacteria can begin to grow there. Pimples appear if the wall of the follicle ruptures, which tells

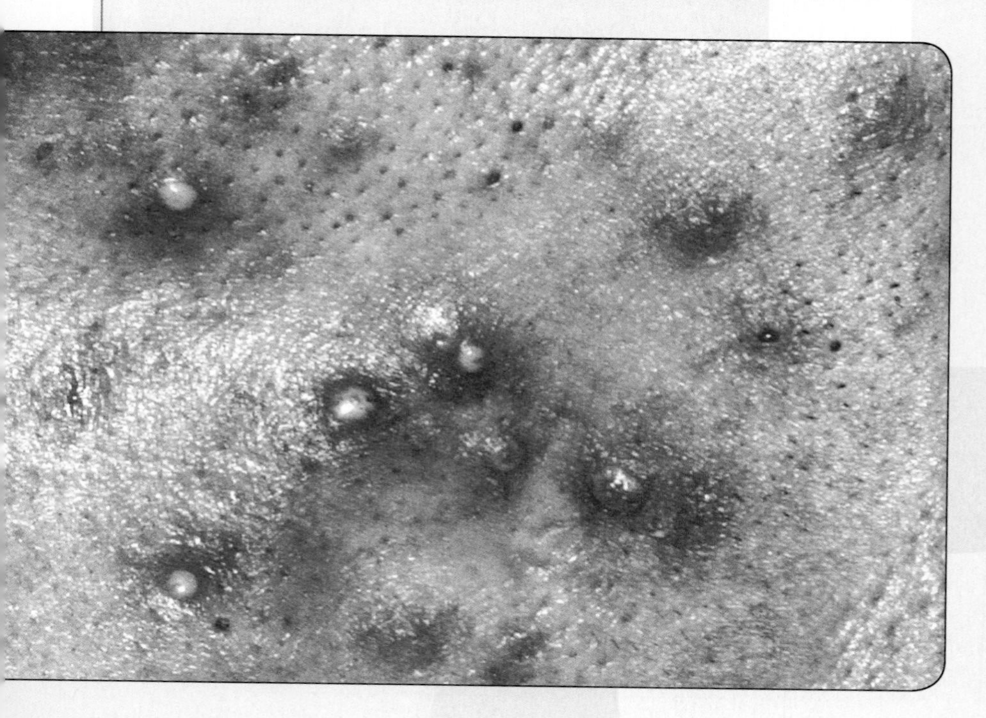

A skin condition like acne is treated by a dermatologist, or skin doctor.

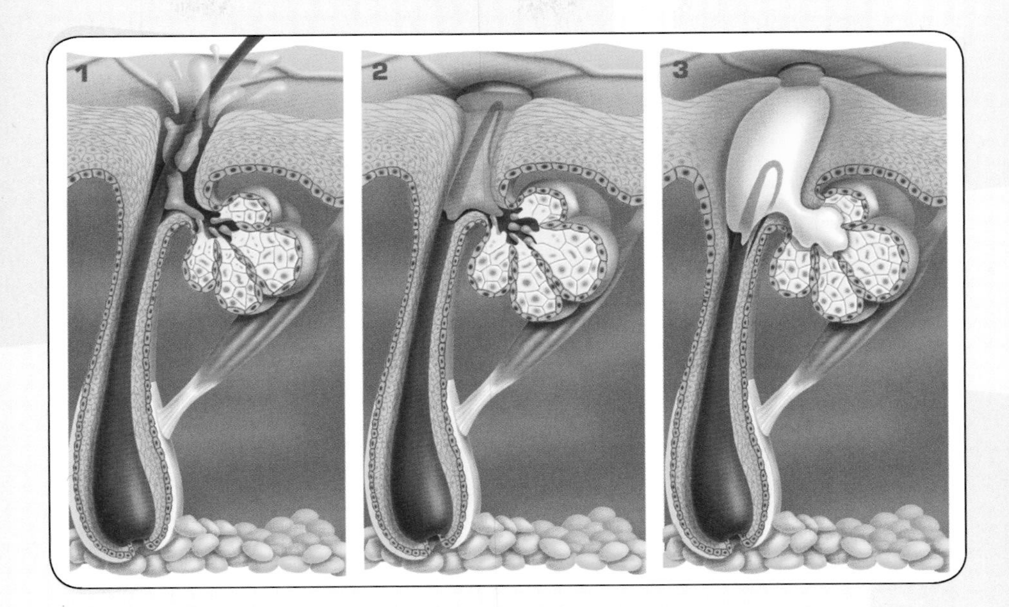

the immune system to fill the pimple with a mucuslike liquid called pus.

There are different kinds of pimples. Whiteheads occur if the blockage stays completely below the skin's surface. If the blockage partially pops out of the skin and becomes exposed to the air, it may turn black from the skin's pigmentation. This is called a blackhead.

Dermatologists believe that the best way to control acne is to keep skin clean. A healthful diet and getting enough sleep and exercise can also help.

(1) Oil is naturally released through a follicle, (2) extra cells mixed with oil clog the follicle, and (3) a pimple forms.

DID YOU KNOW?

The average human head has about 100,000 hairs. You lose about 100 hairs per day from your head.

BONES

Bones are made up of two kinds of tissue—compact bone and spongy bone. Compact bone is a hard layer on the outside of a bone. Underneath the compact bone, there is a layer of spongy bone that contains many holes and air spaces.

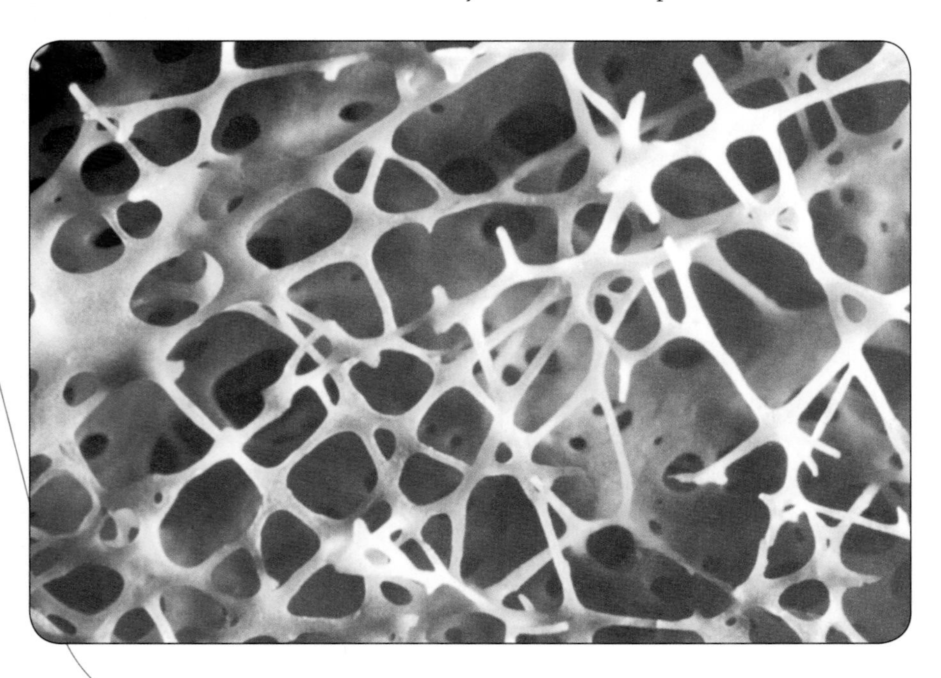

Spongy bone from a human femur (thigh bone) has open spaces that contain bone marrow.

In the center of some bones—including many spongy bone spaces—is a soft tissue called marrow. The ribs, vertebrae, skull, shoulder, and pelvis are a few of the bones that have marrow. Marrow produces red and white blood cells and platelets. Red blood cells transport gases, such as oxygen and carbon dioxide, throughout the body. White blood cells are part of the body's defense system, and platelets help the blood clot.

The primary function of bones is to shape and support the body. Bones also protect softer, fragile organs—such as the heart and lungs—and store minerals like calcium and phosphorous, which help the body function. Bones are essential as a framework for the muscular system.

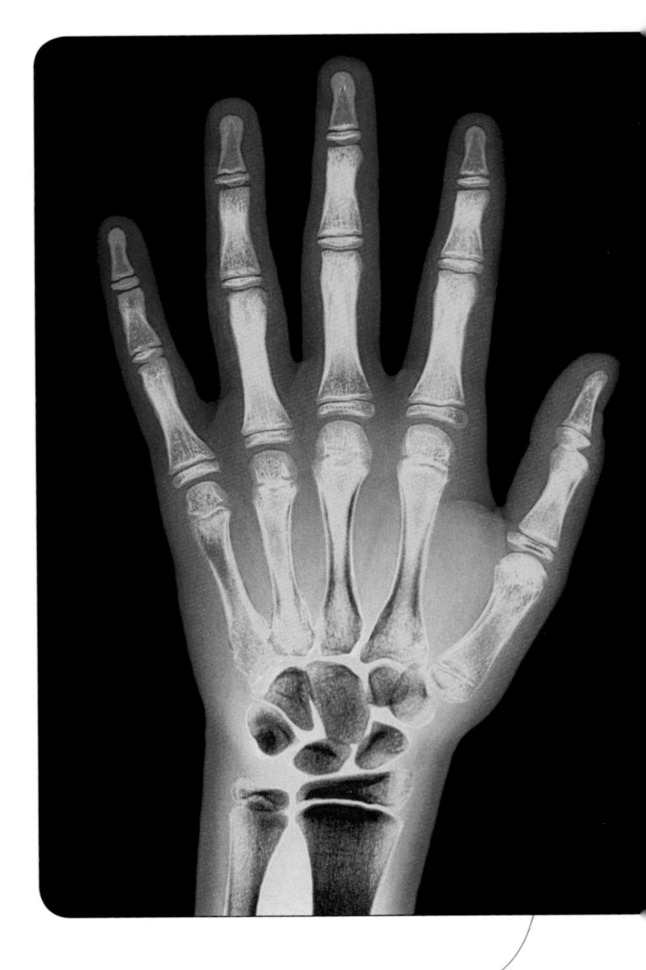

The individual bones in a hand can be seen in an X-ray.

MUSCLES

Muscles are made of groups of long cells called muscle fibers. There are three major types of muscles—smooth, cardiac, and skeletal.

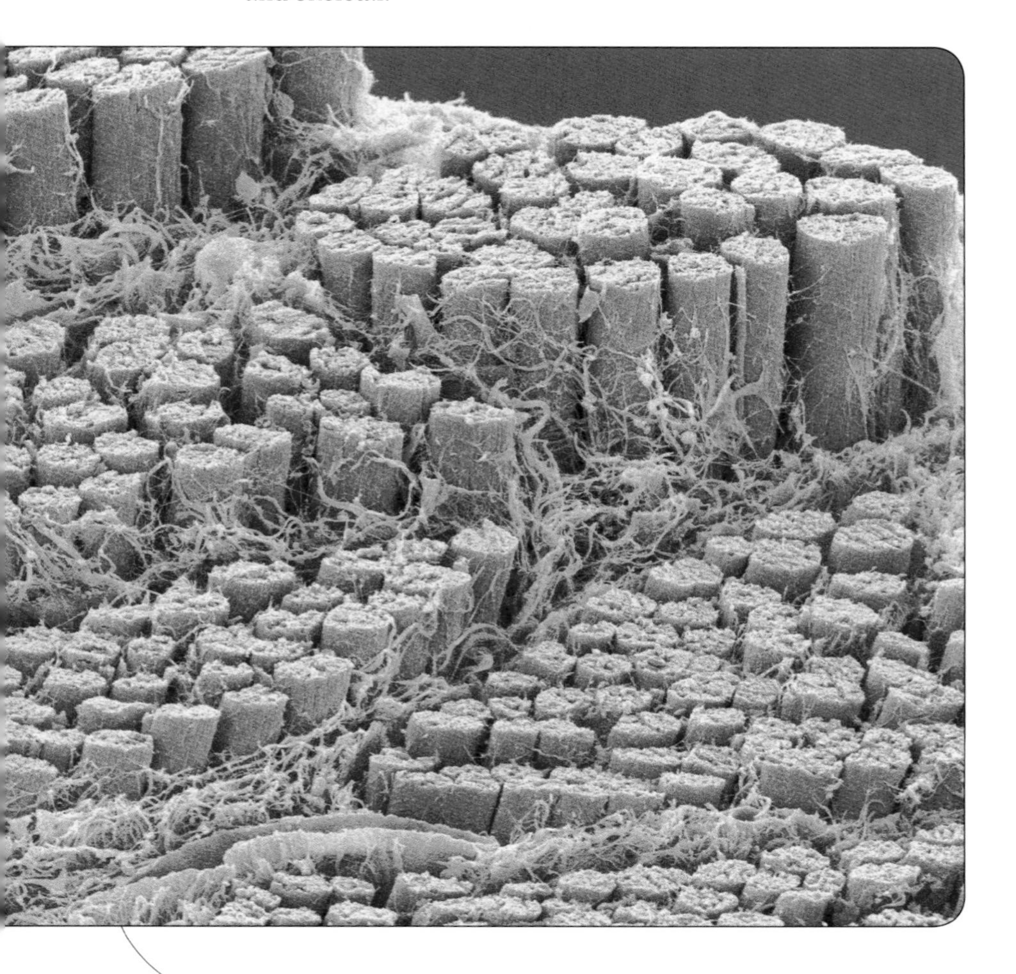

Smooth-muscle fibers contain the proteins actin and myosin, which slide over each other to contract the muscle.

DID YOU KNOW?

The smallest muscle in the body is the stapedius, which is 0.11 inches (0.28 centimeters) long. The stapedius pulls on a tiny bone, the stapes, that sends vibrations from the eardrum to the inner ear.

Smooth muscle is composed of sheets of cells and is found in the walls of hollow internal organs, such as the stomach, intestines, and blood vessels. The body uses smooth-muscle tissue for slow, squeezing actions, such as moving food through the intestine. Smooth muscle moves involuntarily, which means the brain controls the movement of these muscles without any effort from the individual.

The second type of muscle, cardiac muscle, is made of cells that are found in the heart. Cardiac muscle contracts automatically, squeezing the walls of the heart and forcing blood through it. Cardiac-muscle cells work together to help the heart beat in a regular pattern and pump blood efficiently. If cardiac muscle stops working, the person goes into cardiac arrest, which is commonly called a heart attack.

The third type of muscle is skeletal muscle. These muscles are usually attached to bones and joints and help you move. Skeletal muscle is made of fibers that are filled

with bundles of myofibrils. The myofibrils are made of even smaller filaments, which can be thick or thin. The thick filaments are made of a protein called myosin, and the thin filaments are made of another protein called actin. The arrangement of thick and thin filaments give skeletal muscles a striped appearance.

Cardiac muscle can be strengthened through cardiovascular exercises, such as biking, running, and swimming.

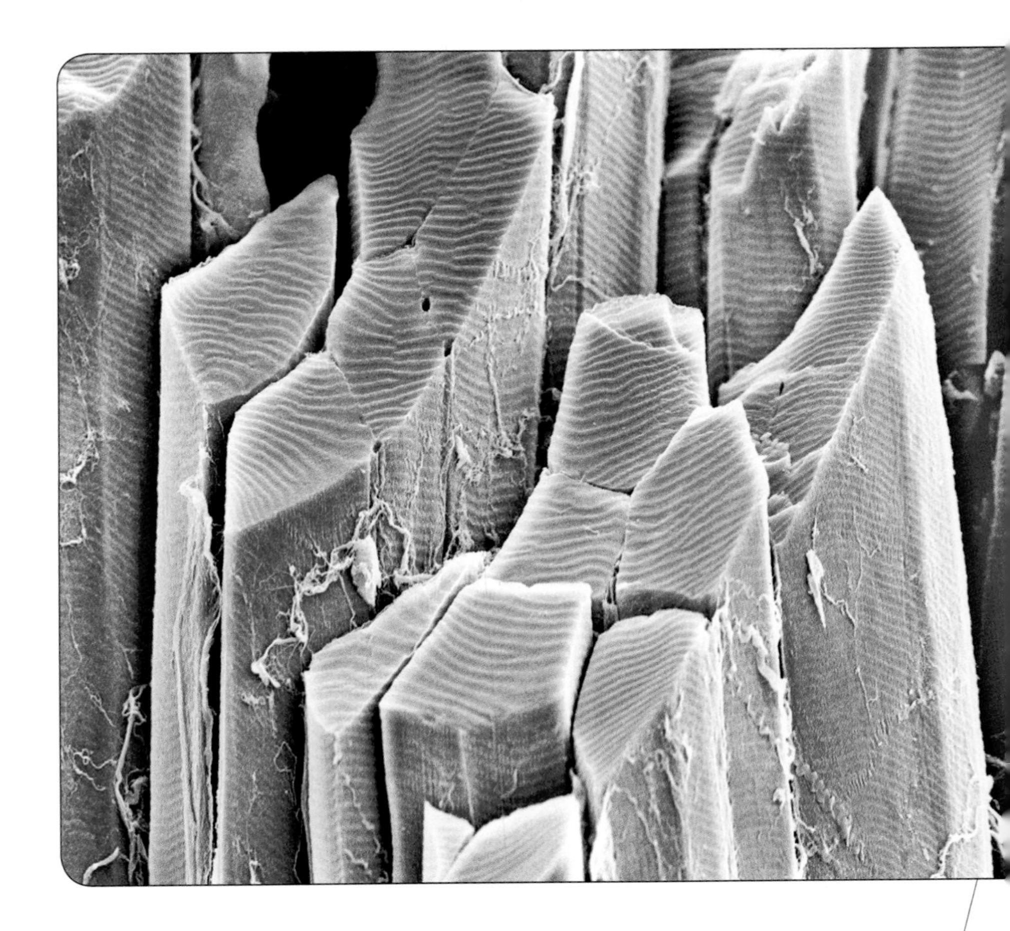

When skeletal muscles shorten, or contract, they pull on the bones to which they are attached. These contractions provide the force and power needed to move the bones. Most skeletal muscles are voluntary muscles because they can be consciously controlled.

Skeletal muscle is also called striated muscle because the arrangement of its proteins has a striped appearance.

CHAPTER THREE

Nervous and Endocrine Organs

THE ORGANS OF the nervous system monitor all of the body's activities. The nervous system is made of two types of systems—the central and peripheral nervous systems. The brain is the major organ of the central nervous system. It is the most complex organ in the body.

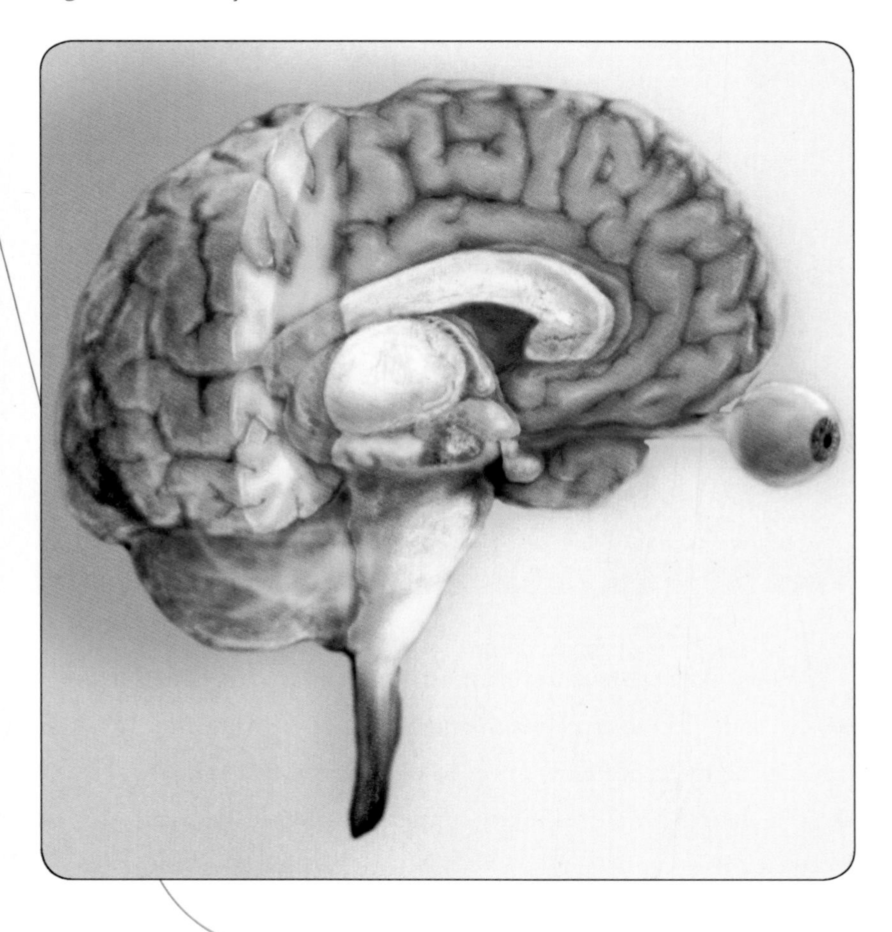

The brain directs all of the functions of the body.

The brain has three main sections: the cerebrum, the cerebellum, and the brain stem. The cerebrum is the largest section of the brain. It is wrinkled and is made of two hemispheres that are connected by nerve tracts. It is the part of the brain that separates us from other animals; it controls intelligence, memory, language, skeletal muscles, and senses. The cerebellum controls balance, posture, and coordination of movements. The brain stem controls the involuntary functions of the body, like breathing and heartbeat.

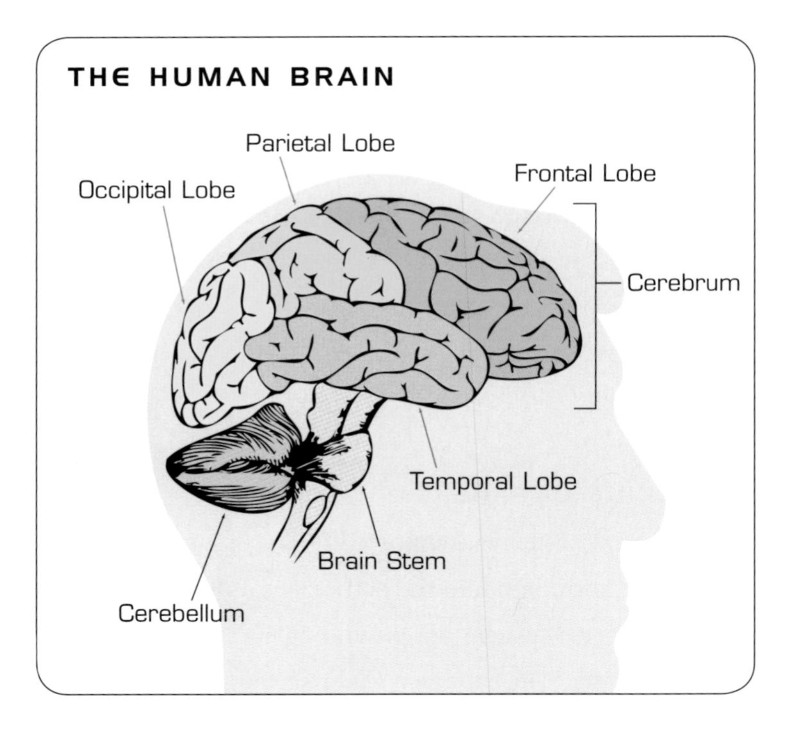

DID YOU KNOW?

Some scientists classify the diencephalon as a main section of the brain. It lies between the two hemispheres, or sides, of the brain and includes the thalamus and hypothalamus. Its functions include maintaining the body's temperature and monitoring appetite and thirst.

The brain connects to the spinal cord, the other major organ of the central nervous system. The spinal cord controls reflexes, or automatic nerve responses, such as when the doctor makes your lower leg kick by hitting your knee with a rubber hammer. The spinal cord is the starting point of 31 pairs of nerves that make up the peripheral nervous system. These nerves provide two-way communication between the central nervous system and the legs, arms, neck, and torso. The skull and vertebrae protect the brain and spinal cord from damage.

ENDOCRINE ORGANS

The endocrine system is composed of cells, tissues, and organs that release hormones into the body. One of the major organs of the endocrine system, the pituitary gland, connects to the nervous system. The pituitary gland secretes important hormones that influence many processes in the body, such as growth, sexual development, metabolism, and reproduction.

The pituitary gland is roughly the size of a pea and is located at the base of the brain. It has an anterior (front) lobe and a posterior (back) lobe. Each lobe receives messages

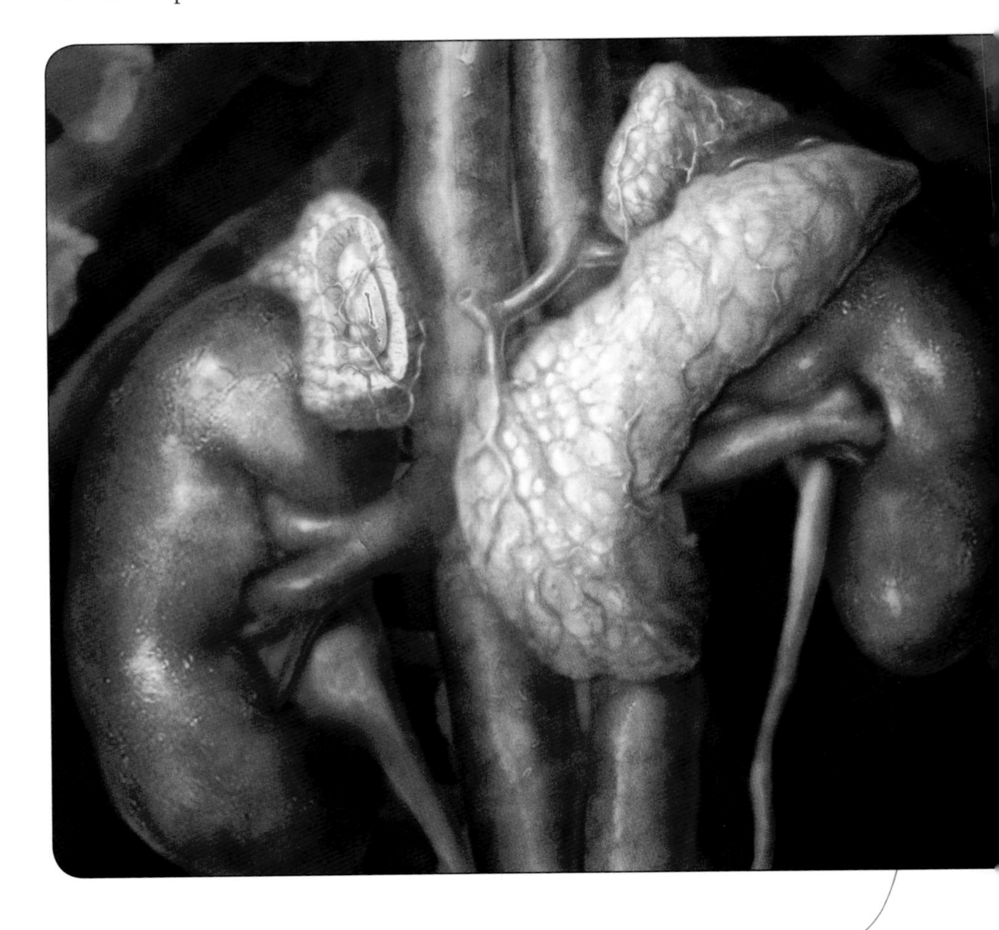

Adrenal glands (white), which are endocrine glands, are perched on top of the kidneys (red).

from the hypothalamus in the brain and sends its hormones through the bloodstream. The pituitary hormones signal various organs to increase or decrease their activities. For example, an increase in certain pituitary hormones signals adrenal glands, two small organs near the kidneys, to speed up or slow down their own hormone production.

Each of the adrenal glands has two parts. The outer region, called the cortex, makes hormones that control the balance of sodium and potassium in the body. It also regulates the level of sugar in the blood and produces sex hormones. The inner region is part of the nervous system. It makes hormones for reacting quickly to fear and anger and is directly connected to the central nervous system.

The thyroid gland, another endocrine organ, is in the neck. The thyroid controls the rate at which the body produces energy from nutrients. The balance of its hormones is delicate. If the thyroid produces too many hormones, the person may lose weight or become restless or emotionally disturbed. If it does not produce enough hormones, bodily functions like digestion may slow down.

The pancreas is a long, thin gland that lies beneath the stomach. Like the adrenals, the pancreas also plays a role in regulating blood sugar. It produces insulin, a hormone that controls the amount of sugar in the blood. In addition

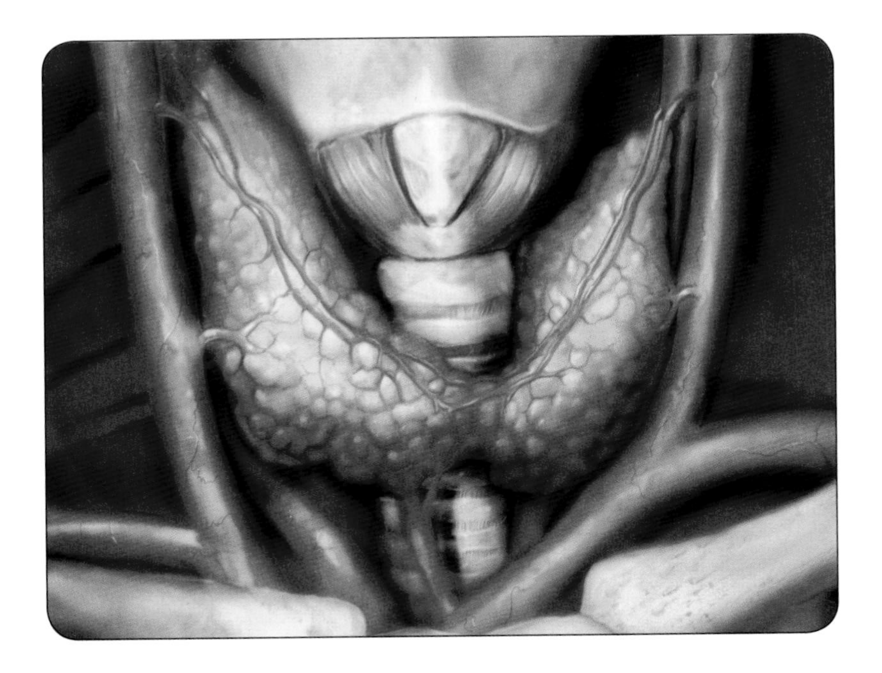

to insulin, the pancreas produces digestive juices that break down fats, carbohydrates, proteins, and acids. It also creates a compound to help neutralize stomach acid as processed food transfers into the intestines.

DID YOU KNOW?

The word *pancreas* comes from a Greek word meaning "all meat." It was given this name because the Greeks considered animal pancreases to be edible.

The thyroid gland (orange) is located near the trachea in the neck.

Diabetes

As part of digestion, carbohydrates and sugars are transformed into glucose to provide energy for all the body's cells. Insulin is a hormone that enables glucose to enter the cells. If the pancreas fails to make enough insulin, the body cannot properly handle sugars, and the glucose level in blood rises to unsafe levels. This is known as type 1 diabetes. Type 1 diabetes can be treated by injections of artificial insulin. This helps the body to process sugar properly.

However, only about 5 percent to 10 percent of people in the United States with diabetes have the type 1 form of the disease. The other 90 percent to 95 percent have type 2 diabetes. This results when the body either does not produce enough insulin or the cells become resistant to the insulin. Type 2 diabetes can also be treated with insulin injections, but exercise and a healthful diet are also highly recommended.

A device called a novopen is used to inject insulin into the bloodstream.

CHAPTER FOUR

Circulatory and Respiratory Organs

WHILE THE NERVOUS and endocrine systems control and coordinate all of the body's activities, they can do nothing without a constant supply of oxygen-rich blood. This is the function of the circulatory system.

The heart is the main organ of the circulatory system. It is a large muscle that contracts and relaxes constantly to keep blood flowing to all parts of the body. Roughly the size of a person's fist, the heart has four chambers. The two upper chambers are called atria, and the two lower chambers are called ventricles. Each time the heart beats, blood

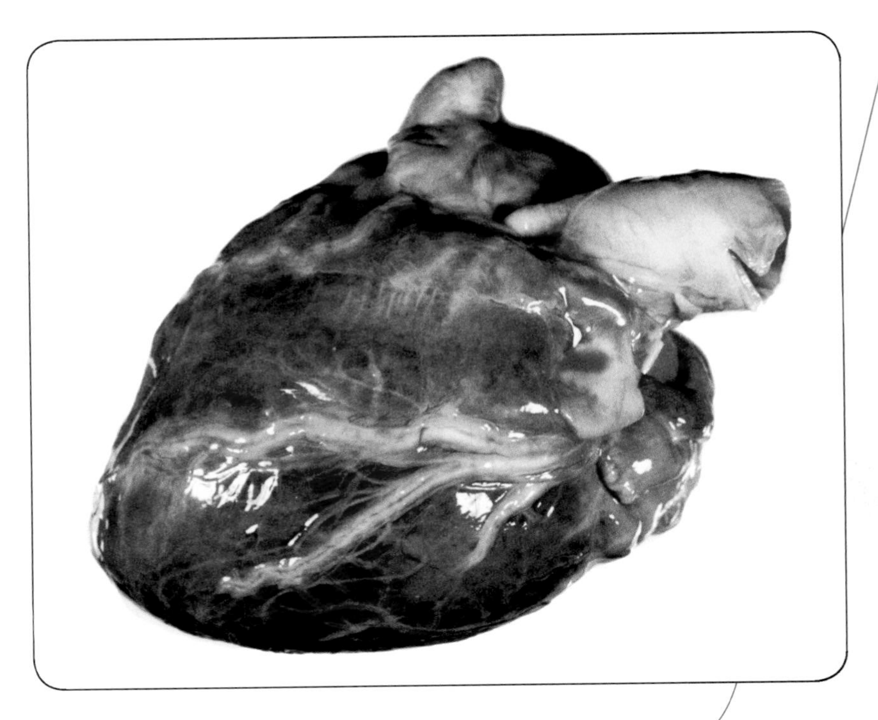

In an average lifetime, the heart beats more than 2.5 billion times.

flows through these chambers.

Blood enters the heart through the atria. Oxygen-poor blood goes into the right atrium, while oxygen-rich blood from the lungs enters the left atrium. Once the atria are filled, they contract and push the blood down into the ventricles. When the right ventricle contracts, it pushes blood out of the heart and toward the lungs through pulmonary arteries—the only arteries that carry oxygen-poor blood. At the same time, the left ventricle pushes its oxygen-rich blood out of the heart through the aorta—the main artery that carries blood from the heart to the rest of the body.

The pumping of the heart causes blood to flow through all the blood vessels and organs of the body. The force of blood on the inner walls of the body's blood vessels is known as blood pressure. In certain elastic blood vessels, you can feel your blood push against and expand the blood vessel wall with each contraction. This is known as your pulse.

There are five types of blood vessels through which blood flows to and from all parts of the body: arteries, arterioles, capillaries, venules, and veins. Arteries are large, muscular tubes that carry oxygen-filled blood away from the heart. The arteries branch into smaller and smaller vessels, called arterioles. The arterioles also branch out until they become capillaries—the narrowest blood vessels in the body. Capillaries are

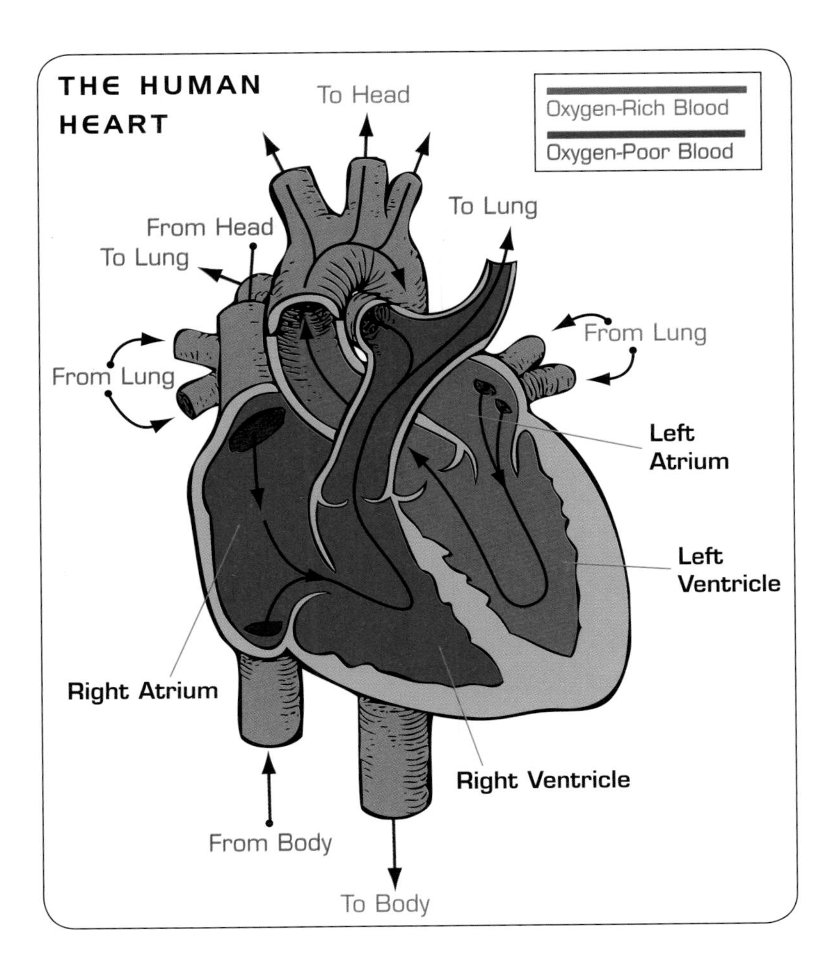

so tiny that they allow nutrients and gases to move from the blood into all the body's tissues.

The blood in the capillaries then flows into venules, which flow into each other as streams flow into rivers. The venules

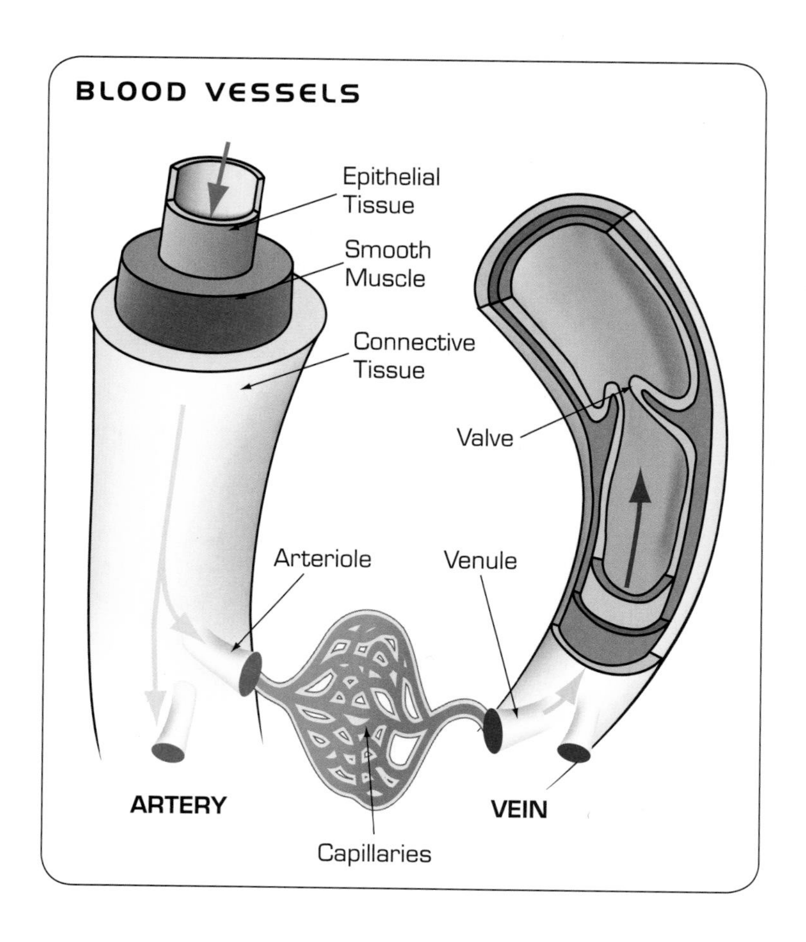

eventually become larger blood vessels called veins, which carry oxygen-poor blood from the tissues and organs back to the heart. The heart then pumps blood to the lungs, where the blood releases carbon dioxide and obtains new oxygen again.

Heart Health

Heart disease is one of the leading causes of death in humans. It can be caused by high blood pressure, abnormal function of the heart's valves, or infection that affects the heart's ability to pump regularly. Heart disease is often caused when arteries become blocked with too much of a fat called plaque. Blockages can develop slowly, over a long period of time. If an artery that feeds the heart muscle becomes mostly or completely blocked, that part of the muscle dies, causing a heart attack.

Sometimes a person is born with heart disease. In many cases, however, heart disease can be easily prevented. Doctors say that regular exercise and a healthful diet can help keep the heart healthy throughout a person's life. People should also avoid harmful activities, such as drug abuse and smoking; the nicotine in cigarettes tightens blood vessels and makes the heart work harder.

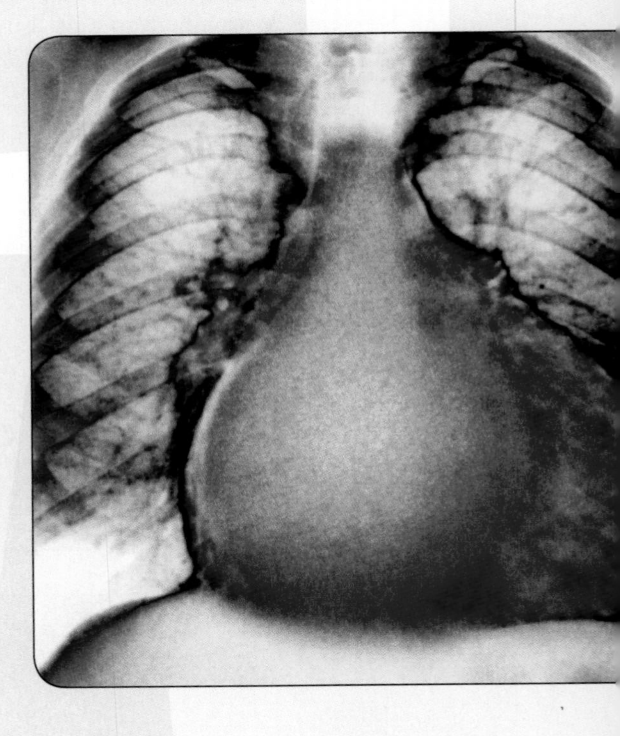

An enlargement of the heart can lead to heart failure, which results in breathlessness and exhaustion. Various drugs may help minimize the symptoms, with heart transplantation being a last resort.

RESPIRATORY ORGANS

The respiratory system brings in new air and disposes of air the body has used. The respiratory system has three major parts—the airways, lungs, and diaphragm. When a person inhales, or breathes in, air passes through the nostrils or mouth and moves down the trachea, a tube that leads to the chest cavity. Here the trachea branches into smaller tubes called bronchi that carry air into the lungs. The trachea and bronchi are filled with tiny hairlike projections called cilia that are covered with mucus. These cilia trap most tiny particles of dirt from the air a person inhales.

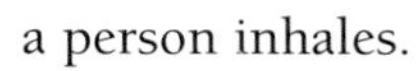

The bronchi of the lungs branch into smaller and narrower tubes, called bronchioles. The bronchioles then branch into even smaller tubules. These tubules lead to alveoli—thousands of tiny sacs with walls that are only a single cell thick. These incredibly thin walls make it possible for oxygen and carbon dioxide to be exchanged. Oxygen-poor

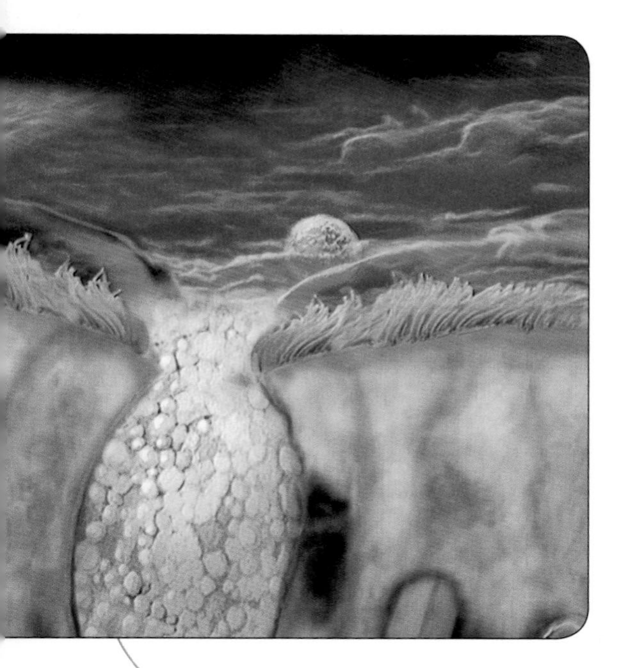

Cilia in the trachea are covered with mucus, which traps invading microbes.

blood from throughout the body enters the alveoli and gives off its carbon dioxide, which is exhaled, or breathed out. At the same time, fresh oxygen moves into the blood. This "refreshed" blood, which is high in oxygen but low in carbon dioxide, exits the lungs and returns to the heart to be pumped back through the body.

Breathing is involuntarily controlled by the brain stem. The diaphragm and muscles between the ribs pull air into the lungs and force it out again. Like little balloons, the alveoli inflate during inhalation and partially deflate during exhalation. When you are excited or exercising, your brain realizes that you need more oxygen and forces you to breathe more quickly. The diaphragm, however, can also be used to voluntarily control breathing when a person holds his or her breath, talks, or plays a musical instrument.

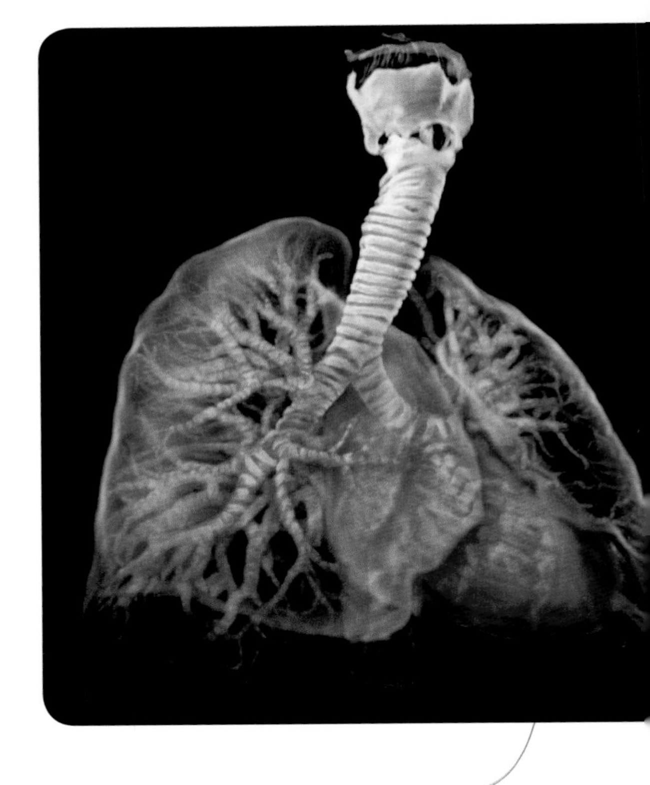

The trachea branches into smaller tubes that aid in respiration.

CHAPTER FIVE

Digestive and Urinary Organs

WHEN YOU FEEL HUNGRY and hear your stomach growling, your digestive system is signaling you to send food its way. Once food has passed through the mouth, it is swallowed and moved into the esophagus, the muscular tube that connects the mouth to the stomach. The muscles of the esophagus go through a series of contrac-
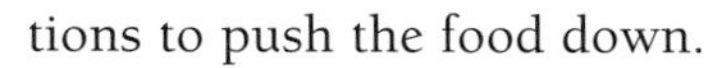
tions to push the food down.

It takes about five to eight seconds for the muscles to move food to the stomach. The stomach is a large, muscular sac that collects the swallowed food. The stomach's function is to further break down the food into particles the body can use as nutrition. The walls of the stomach is made of layers of muscle tissues that cause the food to mix with digestive juices and break apart into chyme.

The lining of the stomach is made of millions of little glands that secrete a combination of highly acidic chemicals called gastric juice. Besides acids, the

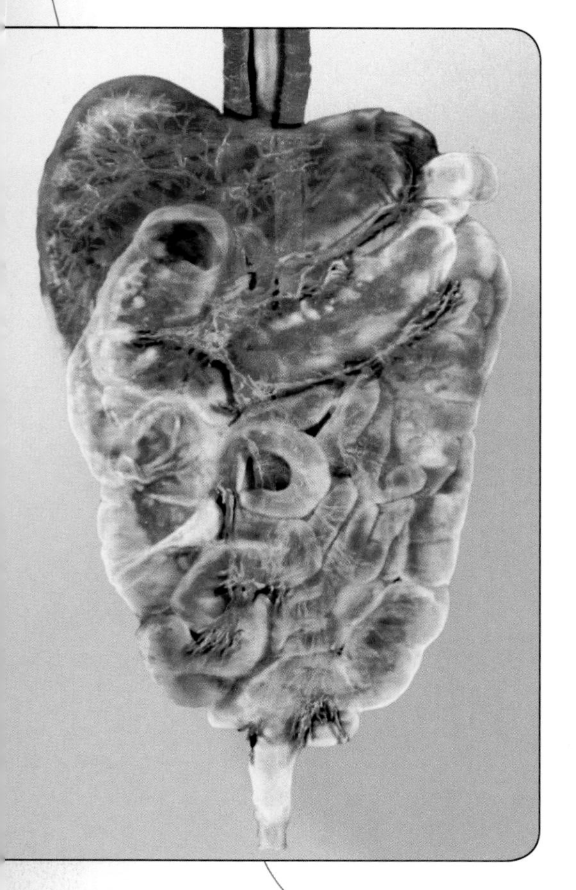

The digestive system includes the stomach, liver, small intestine, and large intestine.

DID YOU KNOW?

A flap of cartilage covers the entrance to the trachea as you swallow. If you happen to laugh or talk while you are swallowing, however, food or liquid might enter the larynx, or windpipe, by accident. This can make you cough.

gastric juice contains pepsin, a substance that breaks down proteins in food.

Stomach acids are quite strong—if they were directly exposed to the lining of the stomach, they would begin to break it down. In order to prevent the stomach from digesting itself, the stomach lining also produces a protective layer of mucus. This mucus prevents acids from having direct contact with the stomach walls and thus limits damage to the organ. Some cells still become damaged by the acids, but these cells are constantly replaced.

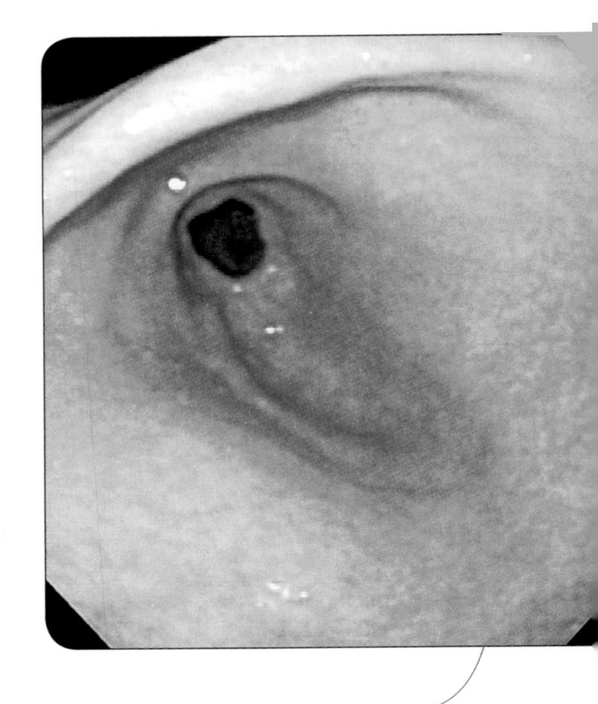

A section of the stomach called the antrum is where most of the churning of the food occurs.

DID YOU KNOW?

The average person eats about 50 tons of food over his or her lifetime.

The liver is a large organ that produces bile, a chemical that breaks down fats and helps neutralize stomach acids. The bile is stored in the gallbladder. From there, the bile and digestive enzymes from the pancreas pass into the small intestine to help with digestion.

The small intestine is a tube of muscle about 20 feet (6 m) long and 1 inch (2.5 cm) in diameter. As food is broken down

The surface of the gallbladder has a rough, lumpy appearance.

into a liquid, the stomach releases small amounts at a time into the small intestine. Chyme is pushed along this tube, where the process of digestion is completed. The carbohydrates, proteins, fats, and acids in food are further broken down by enzymes, which are secreted by the liver, pancreas, and walls of the intestine. Once food is completely digested, most nutrients are absorbed in the small intestine.

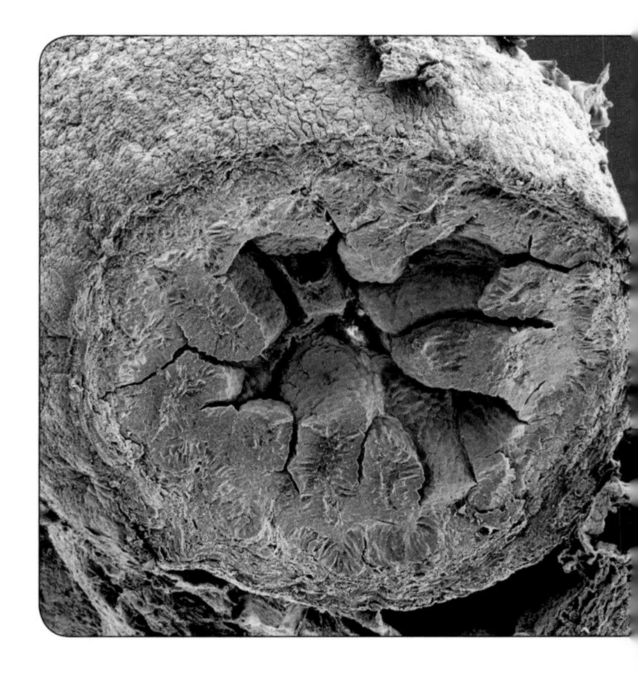

The material that remains passes from the small intestine into the large intestine. The large intestine is a tube that is 5 feet (1.5 m) long and 2.6 inches (6.5 cm) wide, making it shorter and wider than the small intestine. Its job is to absorb any remaining material that the body can use. Its walls absorb water, salts, and some nutrients and vitamins released by special bacteria living in the large intestine. The remaining contents of the large intestine are then eliminated from the body through the anus.

The inside of the small intestine is layered with smooth muscle that contracts, allowing food to pass through smoothly.

URINARY ORGANS

To be efficient and conserve resources, the body recycles and reuses as much as it can. The urinary system is an important part of the body's recycling system. It continually filters fluids, salvaging what it can and disposing the rest in a waste fluid called urine. The kidneys are important parts of this recycling system. About 126 pints (60 liters) of blood enter the kidneys every hour. They filter the blood and reclaim most of the nutrients for reuse. The kidneys reabsorb almost 99 percent of fluids.

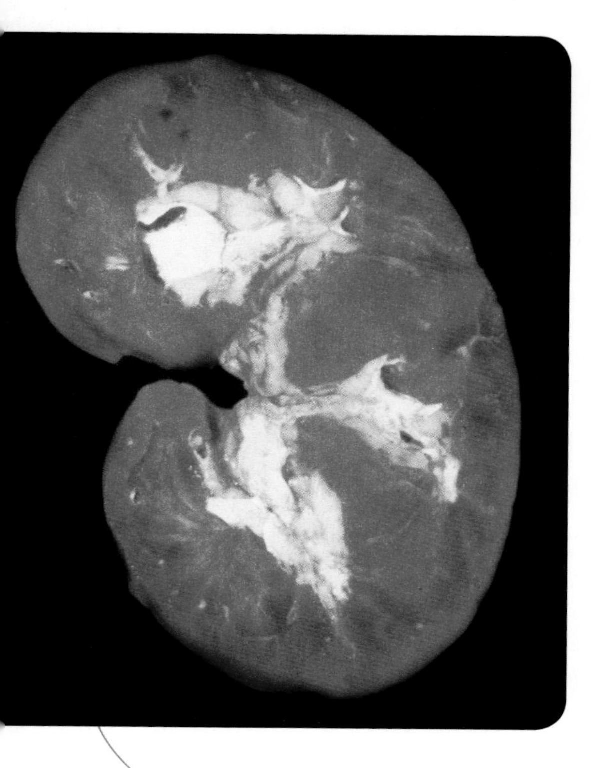

A kidney is approximately the size of a human fist.

Each kidney has about one million tiny filters called nephrons. Blood flows into a nephron onto a large collection of capillary blood vessels. Water, sugars, vitamins, amino acids, salts, and other materials are forced out of the capillaries into a thin-walled sac called a Bowman's capsule. From this capsule, the liquid passes through tiny, narrow tubes where usable materials are reabsorbed into the blood.

Kidneys help to maintain our fluid balance from day to day. For example, if a person drinks a lot one day and sweats extensively the next, the kidneys make sure that the levels of fluids and salts in the blood remain consistent.

Urine moves from the kidneys down 10- to 12-inch (25- to 30-cm) tubes called ureters. The urine moves through the ureters by both gravity and smooth muscle contractions. Each of these tubes attaches to the urinary bladder, which is an expandable sac. The bladder collects the urine and stores it until it can be eliminated from the body. When the bladder gets full, urine exits the body through a small tube called the urethra.

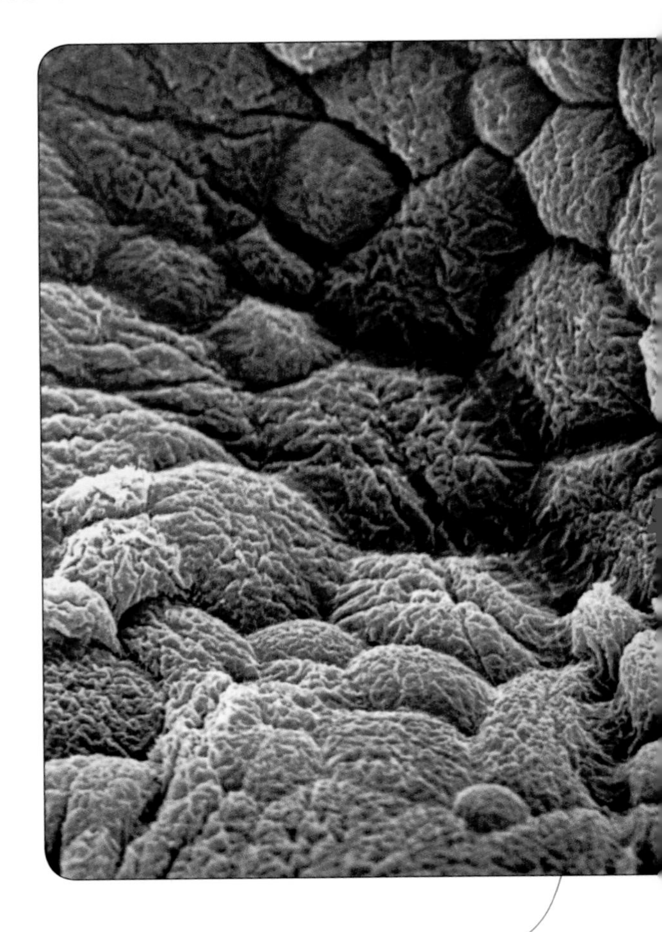

Tightly packed epithelial cells form the lining of the bladder.

Leonardo da Vinci

Although he lived more than 500 years ago, Leonardo da Vinci made many advances in our understanding of the human body. Da Vinci was fascinated by the body's mysteries. He was known to lock himself in his workshop and dissect corpses to discover their parts. Since there were no preservatives back then, he needed to work quickly before the bodies began to decay. He was a careful observer, paying great attention to detail. He drew pictures of the many organs he found and used these drawings to show relationships between what each part looked like and what it did. He studied the brain, bones, muscles, circulatory system, eyes, lungs, and urinary tract. He saw the body as a wonderful machine, and his accomplishments added tremendously to the medical field of his time.

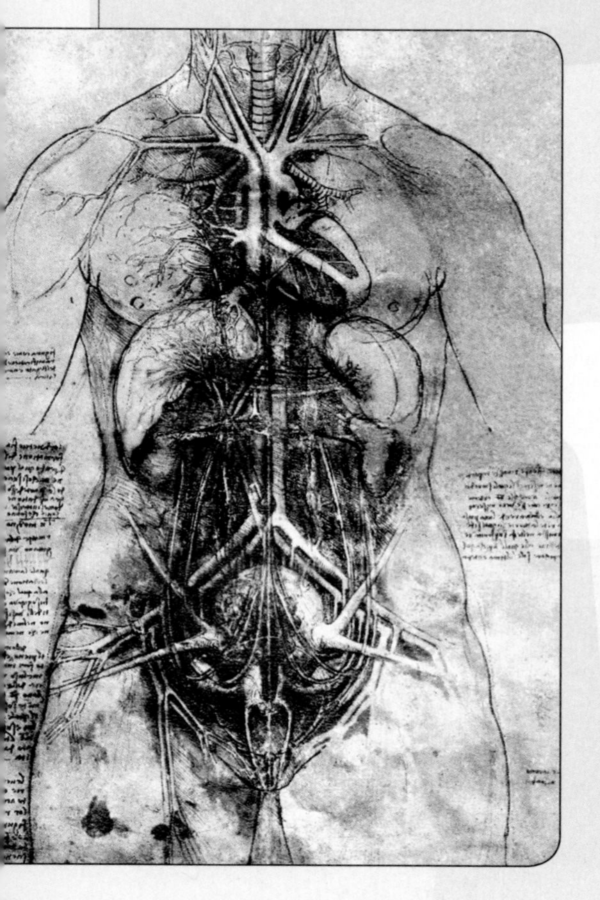

Da Vinci (1452–1519) recorded his findings in detailed diagrams of the human body.

CHAPTER SIX

Reproductive Organs

UNLIKE THE SYSTEMS that focus on body maintenance, the reproductive system's job is to ensure that genes, the body's basic units of heredity, are passed on to the next generation. In other words, the reproductive organs' main purpose is to produce offspring.

Reproductive organs are different in males and females. In males, the main organs are the testes, seminal vesicles, prostate gland, and penis. Testes are found inside the scrotum, a sac of skin that hangs just outside a male's body. One of the main jobs of the testes is to produce sperm, which are the male sex cells.

Inside each testis are tightly coiled tubes where sperm are produced. When the sperm are nearly mature, they move into a tube called the epididymis. Then they move to a larger tube called the vas deferens, where they complete their growth and pass through slowly until they can be released from the body.

The seminal vesicles are a pair of saclike structures found near the base of the urinary bladder in men. They make a fluid that is rich in sugars and provides energy for the sperm cells. The prostate gland is a doughnut-shaped organ that

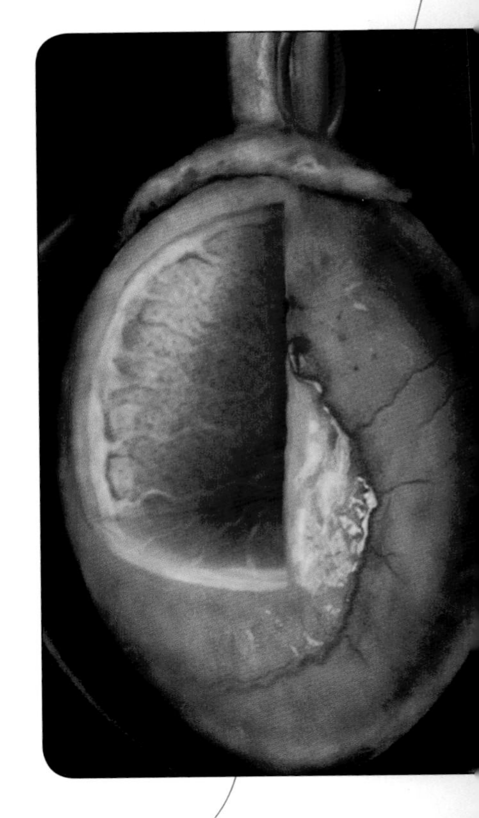

Sperm production sites are located inside the testes.

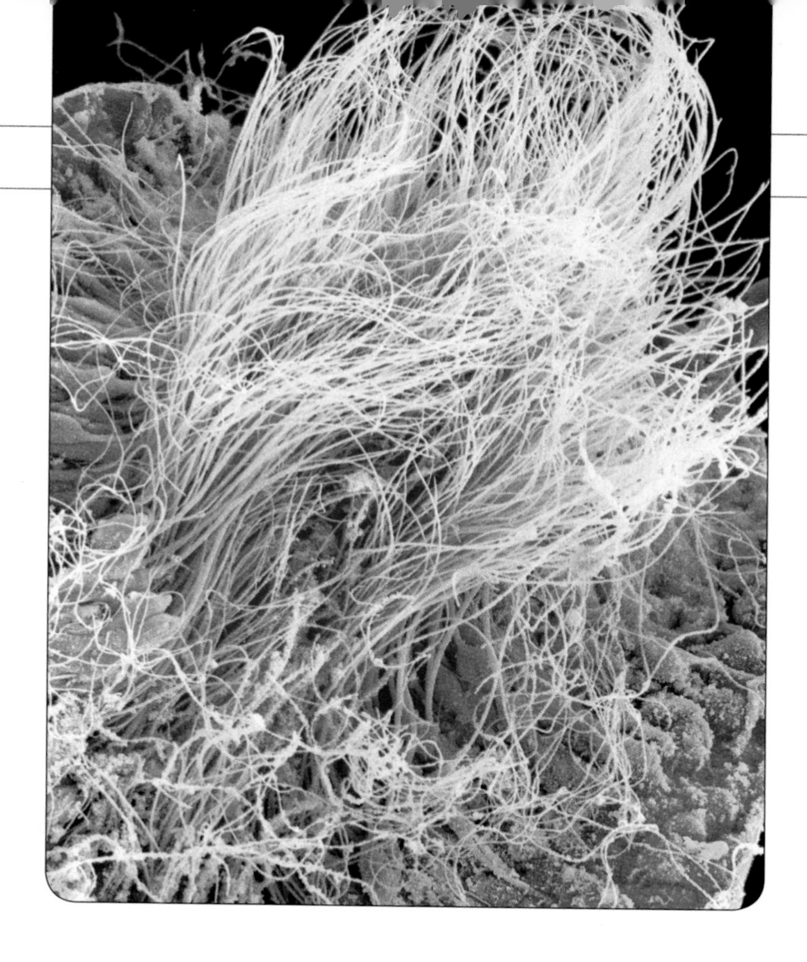

surrounds the urethra. The prostate makes another, thinner fluid that helps sperm swim. When sperm combine with all these fluids, the substance is called semen. Both semen and urine pass through the urethra, which is in the penis. The

DID YOU KNOW?

A single testis can produce as many as 12 trillion sperm in a male's lifetime.

Developing sperm (blue) are nourished by various cells within a sperm-production site.

penis and scrotum make up the male external sex organs.

The internal organs of a female include the ovaries, fallopian tubes, uterus, and vagina. As infants, females carry tens of thousands of eggs—female sex cells. Unlike male organs, which can produce sperm throughout a lifetime, a woman's ovaries carry all of the eggs before she is born. The eggs wait in sacs inside the ovaries, two small organs on either side of the uterus. Ovaries store the eggs and produce two female sex hormones, estrogen and progesterone.

Once a girl reaches puberty, one egg is released from an ovary each month in a process called ovulation. The egg then enters a nearby fallopian tube and moves toward the uterus. At this point, the egg will live for only about 24 hours unless fertilized by a sperm.

If an egg is not fertilized, it dies. Then the uterus sheds its lining in a process called menstruation, and the cycle begins again.

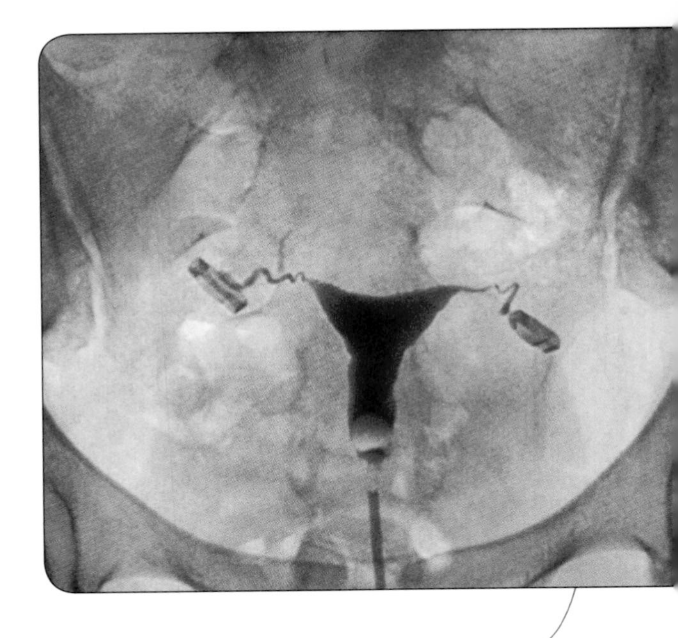

The ovaries, fallopian tubes, and uterus of the female reproductive system can be seen in a colored X-ray.

However, if a sperm finds the egg, it penetrates the egg's wall and fertilizes it in the fallopian tube. Once united, the sperm and egg are collectively called a zygote. The zygote travels through the fallopian tube toward the uterus, dividing many times along the way. By the time it reaches the uterus, it is a hollow ball of cells called a blastocyst. The blastocyst embeds itself in the lining of the uterus.

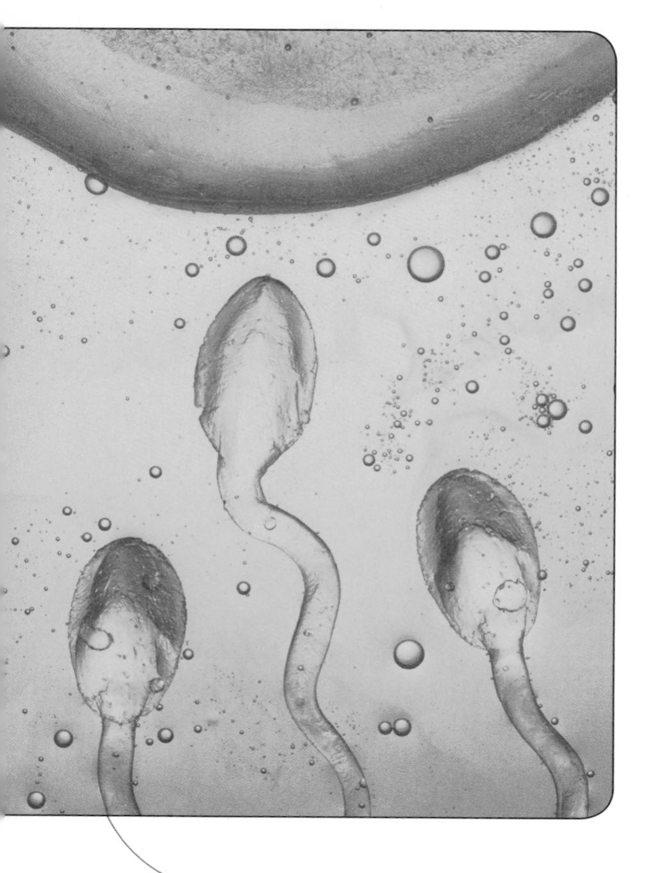

The uterus is a hollow, muscular sac lined with endometrium, a mucus layer that nourishes a developing fetus. The uterus provides the fetus with blood containing nutrients and oxygen. During pregnancy, the uterus grows from a mere 3 to 4 inches (7.5 to 10 cm) across to a size large enough to accommodate a full-grown fetus weighing about 7 pounds (3.2 kilograms).

Sperm use their tails to propel toward the egg.

The Placenta

Females grow an entirely new organ, the placenta, when they are pregnant. The placenta provides a growing fetus with everything it needs to develop properly. The placenta implants in the wall of the uterus and transfers nutrients and oxygen from the female to the fetus. It also transfers fetal wastes to the female's bloodstream for disposal. The placenta also produces hormones that help maintain the health of the female during pregnancy. The placenta is attached to the fetus by the umbilical cord. Once the baby is born, the placenta is no longer needed, and the uterus pushes it out.

The placenta and the fetus develop separately during the first part of the pregnancy, but they are eventually connected by the umbilical cord.

DID YOU KNOW?

The vagina can expand to three times its usual size to allow the head of a baby to pass through.

When a fetus is ready to be born, the uterus's powerful muscles push it out through the vagina, a thin-walled, stretchable passageway that is approximately 3 inches (7.5 cm) in diameter.

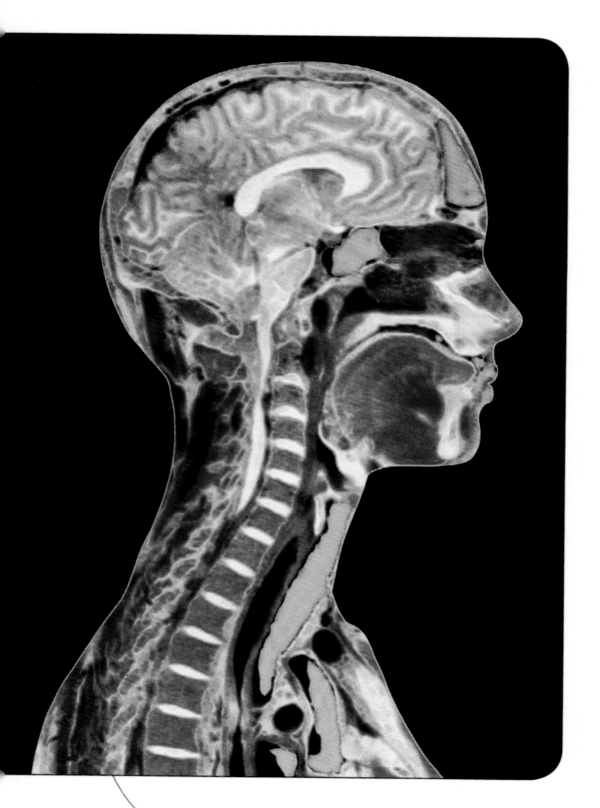

A cross-section of the human body reveals many of the organs that carry out the body's functions.

SUSTAINING LIFE

The human body has many organs, each perfectly adapted for its particular function to keep the body running smoothly. From the brain and stomach to the pituitary gland and liver, each organ knows exactly what to do. All of these organs working together make your body an efficient, living machine.

alveoli—tiny air sacs in the lungs where carbon dioxide and oxygen are exchanged

atria—the two upper chambers of the heart

cerebellum—the part of the brain that coordinates movement

cerebrum—the part of the brain that controls voluntary processes, interprets sensations, and monitors advanced functions like memory and speech

dermatologist—a skin doctor

diencephalon—the part of the brain that includes the thalamus and the hypothalamus

endocrine glands—special organs that produce hormones

endometrium—the inner lining of the uterus

enzymes—special proteins that speed up chemical reactions (such as digestion) in the body

epididymis—the tube where sperm wait to fully mature and be released

fetus—a developing human in its mother's uterus

gall bladder—the organ that sends bile into the small intestine to aid in digestion

hormones—substances that are made in one organ and sent to another

kidneys—two organs that filter fluids in the body and produce urine

liver—the organ responsible for making bile, which aids in digestion

marrow—a soft tissue in the bones that produces red blood cells, white blood cells, and platelets

myofibrils—small units of muscle cells

nephrons—the filtering units of the kidneys

pituitary gland—a small gland that produces important hormones that influence many processes in the body, such as growth, sexual development, metabolism, and reproduction

ventricles—the two lower chambers of the heart

- The human brain weighs about 3 pounds (1.4 kg). That's about 2 percent of the average human body weight.
- The liver is the largest and heaviest internal organ of the body.
- The heart beats about 100,000 times each day.
- Adult humans have 206 bones, but children have about 300. Some bones, especially those in the skull, fuse during the development of the child.
- Human jaw muscles can generate a force of 200 pounds (90 kg) on the molars.
- The human body makes about 7 gallons (27 liters) of digestive juices a day.
- A scab is essentially a bandage of dried blood that the body makes to give the cells underneath a chance to repair themselves without risk of infection.
- The average adult's skin is about 16 to 21.5 square feet (1.5 to 2 square meters) in area, approximately the same area as a twin-size bed.

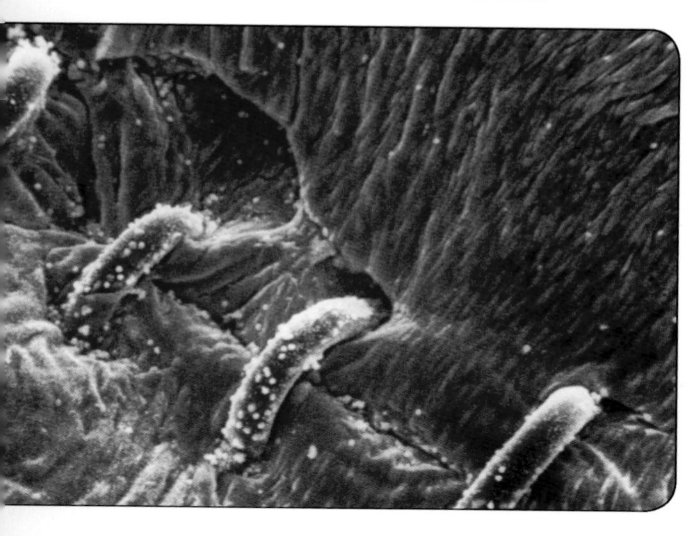

- Doctors use string made of nylon, silk, or other materials to reconnect skin that has been cut deeply. They use loops of string and tiny knots to hold the skin together until it heals. These loops are called stitches. Once the wound heals, the stitches are removed. Some stitches will dissolve once the wound heals.

Stitches and cut skin can be seen in fine detail through the use of a scanning electron micrograph (SEM).

At the Library

Cassan, A. *The Brain.* Philadelphia: Chelsea House, 2006.
Gray, Susan H. *The Heart.* Chanhassen, Minn.: Child's World, 2006.
Newquist, H.P. *The Great Brain Book: An Inside Look at the Inside of Your Head.* New York: Scholastic Reference, 2004.

On the Web

For more information on this topic, use FactHound.

1. Go to *www.facthound.com*
2. Type in this book ID: 0756519594
3. Click on the *Fetch It* button.

FactHound will find the best Web sites for you.

On the Road

Chicago Museum of Science and Industry
57th Street and
Lake Shore Dr.
Chicago, IL 60637
773/684-1414

The John P. McGovern Museum of Health and Medical Science
1515 Hermann Dr.
Houston, TX 77004
713/521-1515

Explore all the Life Science books

Animal Cells: The Smallest Units of Life

DNA: The Master Molecule of Life

Genetics: A Living Blueprint

Human Body Systems: Maintaining the Body's Functions

Major Organs: Sustaining Life

Plant Cells: The Building Blocks of Plants

A complete list of Exploring Science titles is available on our Web site: *www.compasspointbooks.com*